Solar Eclipse 2024

Witnessing the Celestial Phenomenon of April 8 and Exploring Its Science, Myths, and Cultural Significance

By Mark M. Fields

Copyright © 2024 by Mark M. Fields

All rights reserved. No part of this book may be reproduced, distributed, or transmitted in any form or by any means, including photocopying, recording, or other electronic or mechanical methods, without the prior written permission of the publisher, except in the case of brief quotations embodied in critical reviews and certain other noncommercial uses permitted by copyright law.

Disclaimer

The information provided in this book is for educational and informational purposes only. While every effort has been made to ensure accuracy, the author and publisher make no representations or warranties with respect to the accuracy or completeness of the contents of this book and specifically disclaim any implied warranties of merchantability or fitness for a particular purpose. The author and publisher shall not be liable for any loss of profit or any other commercial damages, including but not limited to special, incidental, consequential, or other damages. Readers are encouraged to consult with qualified professionals for specific advice tailored to their individual circumstances.

About the Author

Mark M. Fields is an avid astronomer, science enthusiast, and dedicated educator with a passion for sharing the wonders of the universe with others. With a background in astrophysics and a lifelong fascination with celestial phenomena, Mark has devoted his career to exploring the mysteries of space and inspiring curiosity about the cosmos.

As an experienced educator, Mark has taught astronomy and science courses at both the secondary and post-secondary levels, engaging students of all ages in hands-on learning experiences and exciting scientific discoveries. He believes in the power of education to ignite curiosity, foster critical thinking, and inspire the next generation of scientists and explorers.

In addition to his work in education, Mark is an active member of the astronomy community, participating in research projects, outreach events, and amateur observing programs. He has traveled to numerous eclipse events around the world, capturing stunning photographs and sharing his expertise with fellow eclipse enthusiasts.

With "Solar Eclipse 2024," Mark combines his passion for astronomy, his dedication to education, and his commitment to sharing the wonders of eclipses with readers everywhere. Through this book, he hopes to inspire awe, curiosity, and a deeper appreciation for the beauty and complexity of the universe.

Table of Contents

Introduction..9
    Explanation of What a Solar Eclipse Is............. 9
    Importance and Fascination of Solar Eclipses 11
The Science of Solar Eclipses............................ 15
    How Solar Eclipses Occur................................15
    Types of Solar Eclipses....................................17
    Path of Totality and Viewing Locations............22
History of Solar Eclipses.....................................27
    Ancient Observations and Interpretations.......27
    Contributions to Scientific Understanding....... 32
    Notable Historical Eclipses..............................37
Cultural Significance.. 43
    Myths, Legends, and Superstitions Surrounding Eclipses..43
    Influence on Art, Literature, and Folklore........ 47
    Eclipse Rituals and Traditions from Around the World...52
Eclipse Viewing Safety... 57
    Importance of Eye Protection During Solar Eclipses..57
    Safe Viewing Methods and Equipment for Solar Eclipses..61
    Health risks of viewing eclipses improperly.....65
Planning for an Eclipse.. 71
    How to prepare for an eclipse viewing............ 71
    Choosing a viewing location........................... 76
    Weather considerations.................................. 80
Experiencing a Solar Eclipse............................87

    What to expect during each phase of the eclipse.............87
    Emotional and psychological impact of totality.............92
    Capturing eclipse phenomena through photography.............96

Beyond the Eclipse.............103
    Scientific research during eclipses.............103
    Citizen science opportunities.............108
    Future eclipse predictions and events.............113

Frequently Asked Questions.............117
    common misconceptions and concerns.............117
    Answers to common questions about eclipses.............122

Thanks! Readers.............127

Introduction

Explanation of What a Solar Eclipse Is

A solar eclipse is a breathtaking celestial event that occurs when the Moon passes between the Earth and the Sun, temporarily blocking the Sun's light from reaching the Earth's surface. This captivating phenomenon can only transpire during the New Moon phase when the Sun, Moon, and Earth are perfectly aligned in syzygy.

During a solar eclipse, the Moon casts its shadow on the Earth, resulting in either a partial or total obscuration of the Sun, depending on one's location within the eclipse's path. There are three primary types of solar eclipses:

1. **Total Solar Eclipse:** When the Moon completely covers the Sun, revealing the Sun's ethereal corona, the outermost layer of the Sun's atmosphere. This remarkable event plunges the area within the Moon's shadow, known as the path of totality, into an eerie darkness akin to twilight.
2. **Partial Solar Eclipse:** In regions outside the path of totality, observers witness a partial solar eclipse, where only a portion of the Sun is obscured by the Moon. This results in a crescent-shaped Sun as seen from Earth.
3. **Annular Solar Eclipse:** An annular solar eclipse occurs when the Moon is farther away from Earth in its elliptical orbit, causing it to appear slightly smaller than the Sun. As a result, during the maximum phase of

the eclipse, the Moon does not completely cover the Sun, leaving a dazzling ring of sunlight, or "annulus," around the Moon's silhouette.

Solar eclipses are rare and awe-inspiring events, captivating the imagination of humanity for millennia. They serve as a reminder of the cosmic dance of celestial bodies and our place in the vast universe. The anticipation and excitement surrounding solar eclipses often bring people together to witness the wonder of nature firsthand.

Importance and Fascination of Solar Eclipses

Solar eclipses have captivated human imagination since antiquity, playing a significant role in cultures, scientific discovery, and personal experiences. Their importance and fascination stem from various factors:

1. **Scientific Inquiry:** Solar eclipses provide invaluable opportunities for scientific research and discovery. During a total solar eclipse, the Sun's outer atmosphere, known as the corona, becomes visible to the naked eye, offering scientists a rare chance to study its structure, temperature, and dynamics. Eclipses also allow astronomers to test theories, validate models, and advance our understanding of celestial mechanics.
2. **Cultural Significance:** Across cultures and civilizations, solar eclipses have been imbued with deep cultural and religious significance. Ancient civilizations often interpreted eclipses as omens or messages from the gods, influencing religious rituals, myths, and folklore. Today, solar eclipses continue to inspire awe and wonder, fostering a sense of

connection to the cosmos and the passage of time.

3. **Unity and Community:** Solar eclipses have a unique ability to bring people together, transcending boundaries of geography, language, and culture. Communities around the world gather to witness these rare events, fostering a sense of unity and shared experience. Whether observing from bustling city streets or remote mountaintops, eclipse viewers share in the awe and excitement of witnessing nature's grand spectacle.

4. **Personal Experience:** For individuals fortunate enough to witness a total solar eclipse, the experience is often described as transformative and unforgettable. The sudden darkness, the eerie silence, and the sight of the Sun's corona evoke powerful emotions and stir a profound sense of wonder. Many eclipse chasers travel thousands of miles to stand in the path of totality, seeking to immerse

themselves in this once-in-a-lifetime experience.
5. **Inspiration for Creativity:** Solar eclipses have inspired artists, writers, and musicians throughout history, serving as motifs in literature, paintings, and compositions. From ancient cave paintings to modern-day photography and film, eclipses continue to spark creativity and imagination, inviting interpretation and expression across artistic mediums.

In summary, the importance and fascination of solar eclipses lie in their scientific significance, cultural resonance, ability to foster unity, personal impact on observers, and inspiration for creativity. As we look forward to future eclipses, we are reminded of the enduring wonder and mystery of the cosmos, inviting us to explore, learn, and marvel at the wonders of the universe.

The Science of Solar Eclipses

How Solar Eclipses Occur

The mesmerizing phenomenon of a solar eclipse occurs when the Moon, Earth, and Sun align in a precise configuration, resulting in the Moon casting its shadow on the Earth's surface. Understanding the mechanics behind solar eclipses requires insight into the orbits and motions of these celestial bodies.

1. **The Earth's Orbit:** Earth orbits the Sun in an elliptical path, completing one orbit approximately every 365.25 days. This orbit is slightly tilted relative to the plane of the solar system, known as the ecliptic plane.
2. **The Moon's Orbit:** The Moon orbits the Earth in an elliptical path, completing one orbit approximately every 27.3 days. The Moon's orbit is also slightly inclined relative to the ecliptic plane, with its path intersecting the plane at two points called nodes.
3. **Alignment of the Sun, Earth, and Moon**: A solar eclipse occurs during the New Moon phase when the Moon is positioned between the Earth and the Sun, forming a straight line known as syzygy. However, not every New Moon results in a solar eclipse because of the Moon's orbital inclination.
4. **Duration and Frequency:** Solar eclipses are relatively brief events,

16

with the total phase of a total solar eclipse lasting only a few minutes at any given location. However, the entire eclipse, from partial phase to totality and back, can last several hours. Solar eclipses occur several times a year but are only visible from specific locations on Earth due to the narrow path of totality.

Understanding the intricate dance of the Sun, Earth, and Moon is key to comprehending the occurrence of solar eclipses. These celestial alignments create moments of wonder and awe as we witness the interplay of light and shadow in the vast expanse of space.

Types of Solar Eclipses

Solar eclipses manifest in different forms depending on the alignment of the

Sun, Moon, and Earth, resulting in three primary types: total, partial, and annular. Each type offers a unique spectacle, captivating observers with its distinctive characteristics.

1. **Total Solar Eclipse:**
    - **Description**: A total solar eclipse occurs when the Moon completely covers the Sun, obscuring its entire disk from view. This alignment creates a stunning visual spectacle, with the Sun's ethereal corona—a halo of plasma surrounding the Sun's surface—visible to the naked eye.
    - **Path of Totality:** Totality, the phase when the Sun is completely obscured, occurs along a narrow path on the Earth's surface. Observers within this path experience a sudden transformation as daylight gives way to darkness,

revealing stars and planets in the daytime sky.
- **Duration**: The duration of totality varies but typically lasts only a few minutes at any given location. Despite its brief duration, the experience of witnessing a total solar eclipse is profoundly moving and unforgettable.

2. **Partial Solar Eclipse:**
   - **Description**: In a partial solar eclipse, the alignment of the Sun, Moon, and Earth results in only a portion of the Sun being obscured by the Moon. From the perspective of an observer on Earth, the Sun appears as a crescent, with a portion of its disk still visible.
   - **Visibility**: Partial solar eclipses are visible over a much broader geographic area compared to total eclipses, as the Moon's

shadow partially covers the Sun across a wider region.
- **Duration**: The duration of a partial solar eclipse varies depending on one's location within the eclipse's path. Observers may experience partial coverage for an extended period, from the beginning of the eclipse until its conclusion.

3. **Annular Solar Eclipse:**
    - **Description**: An annular solar eclipse occurs when the Moon is at a relatively distant point in its orbit, causing it to appear slightly smaller than the Sun. As a result, during the maximum phase of the eclipse, the Moon does not completely cover the Sun's disk. Instead, a bright ring, known as an annulus or "ring of fire," surrounds the dark silhouette of the Moon.

- **Ring of Fire:** The annular phase of the eclipse creates a mesmerizing sight, with the Sun's outer edges visible as a brilliant ring encircling the Moon's silhouette. Despite the ring of sunlight, observers do not experience the same level of darkness as during a total solar eclipse.
- **Visibility**: Annular eclipses are visible from specific regions along the eclipse's path, offering a unique spectacle to observers within the path of annularity.

Understanding the distinctions between total, partial, and annular solar eclipses allows observers to appreciate the diversity of celestial phenomena and the remarkable interplay of light and shadow in the cosmos. Whether witnessing the awe-inspiring beauty of totality or the subtle crescent phases of a partial eclipse, each type of solar eclipse offers a mesmerizing glimpse into the cosmic ballet of celestial bodies.

Path of Totality and Viewing Locations

The path of totality is a narrow corridor on the Earth's surface where observers can experience the awe-inspiring spectacle of a total solar eclipse, with the Sun completely obscured by the Moon. Understanding the path of totality and selecting optimal viewing locations are crucial for witnessing this rare and breathtaking event.

1. **Definition of the Path of Totality:**
    - The path of totality is the track traced by the Moon's shadow as it sweeps across the Earth during a total solar eclipse. Within this narrow corridor,

observers have the opportunity to witness the complete obscuration of the Sun, leading to a temporary darkness akin to twilight.

2. **Width and Duration:**
    - The width of the path of totality varies depending on the geometry of the Sun, Moon, and Earth during the eclipse. Typically, it ranges from a few dozen to a few hundred kilometers wide.
    - The duration of totality also varies along the path, with the maximum duration occurring at the centerline of the path and decreasing gradually towards the edges.
3. **Selecting Optimal Viewing Locations:**
    - When selecting a viewing location within the path of totality, several factors should be considered:

- **Accessibility**: Choose a location that is easily accessible and allows for a clear view of the sky, free from obstructions such as buildings or trees.
- **Weather**: Check the weather forecast for potential cloud cover or precipitation, as clear skies are essential for optimal viewing.
- **Duration of Totality**: Consider the duration of totality at your chosen location, as longer durations offer more time to observe and appreciate the eclipse.
- **Amenities**: Plan ahead for amenities such as restrooms, food, and water, especially if you

are traveling to remote viewing locations.
4. **Popular Viewing Locations:**
    - Several regions and cities along the path of totality are popular destinations for eclipse enthusiasts. Some notable viewing locations for the upcoming total solar eclipse on April 8, 2024, include:
        - Dallas, Texas
        - Indianapolis, Indiana
        - Buffalo, New York
        - Burlington, Vermont
        - Montreal, Quebec (Canada)
5. **Travel Considerations:**
    - If traveling to witness the eclipse, plan your accommodations and transportation well in advance, as hotels and rental cars may book up quickly.
    - Be prepared for traffic congestion, especially on the

day of the eclipse, as thousands of people may converge on viewing locations within the path of totality.
- Arrive at your chosen viewing location early to secure a prime spot and avoid last-minute crowds.

By understanding the path of totality and carefully selecting optimal viewing locations, eclipse enthusiasts can maximize their chances of experiencing the awe and wonder of a total solar eclipse firsthand. Whether observing from bustling city streets or remote countryside, witnessing the celestial spectacle of totality is an unforgettable experience that transcends boundaries and unites people in awe of the cosmos.

History of Solar Eclipses

Ancient Observations and Interpretations

Solar eclipses have captured the human imagination since ancient times, inspiring awe, fear, and wonder among civilizations around the world. Ancient cultures observed these celestial events and developed diverse interpretations and beliefs to explain their occurrence. Here are some examples of ancient observations and interpretations of solar eclipses:

1. **Mesopotamia and Ancient Near East:**
    - In Mesopotamia, one of the cradles of civilization, solar

eclipses were often interpreted as omens or signs from the gods. Ancient Mesopotamian texts, such as the Enuma Anu Enlil, record observations of eclipses and their perceived significance. Eclipses were seen as messages from the gods, signaling impending changes or divine displeasure.

2. **Ancient Egypt**:
    - In ancient Egypt, solar eclipses were linked to the activities of celestial deities, particularly the sun god Ra. Eclipses were viewed as moments when Ra battled with cosmic forces or encountered challenges on his daily journey across the sky. Rituals and offerings were performed to appease the gods and ensure the sun's return.
3. **Ancient China:**
    - Ancient Chinese astronomers meticulously recorded solar

eclipses and developed sophisticated methods for predicting their occurrence. The Chinese viewed eclipses as harbingers of dynastic changes or natural disasters. Eclipse observations were carefully documented in historical texts such as the "Bamboo Annals" and the "Records of the Grand Historian."

4. **Ancient Greece and Rome:**
   - In ancient Greece and Rome, eclipses were often interpreted as supernatural events, with various myths and legends associated with their occurrence. Philosophers such as Aristotle and Ptolemy sought to understand the scientific principles behind eclipses, laying the groundwork for early theories of celestial motion.

5. **Indigenous Cultures**:
   - Indigenous cultures around the world developed their own interpretations of solar eclipses based on their cosmological beliefs and cultural traditions. For example, among the indigenous peoples of North America, eclipses were sometimes seen as symbolic of cosmic balance or as reminders of the interconnectedness of all living beings.
6. **Maya Civilization:**
   - The ancient Maya civilization of Mesoamerica possessed advanced astronomical knowledge and developed precise calendars to track celestial events, including solar eclipses. Mayan hieroglyphic inscriptions and codices contain references to eclipses and their significance in religious rituals and ceremonies.

7. **Ancient India:**
    - In ancient India, solar eclipses were viewed as celestial phenomena with both astronomical and spiritual significance. Hindu texts such as the Vedas and Puranas contain references to eclipses and prescribe rituals and observances to mitigate their effects and ensure auspicious outcomes.

Ancient observations and interpretations of solar eclipses reflect humanity's enduring fascination with the cosmos and our efforts to understand the mysteries of the universe. These early interpretations laid the foundation for modern scientific inquiry into eclipses, contributing to our evolving understanding of celestial mechanics and the interconnectedness of the natural world.

Contributions to Scientific Understanding

Solar eclipses have played a crucial role in advancing our scientific understanding of the universe, serving as catalysts for groundbreaking discoveries and observations. Throughout history, astronomers and scientists have leveraged solar eclipses to study phenomena ranging from the structure of the Sun's atmosphere to the validation of Einstein's theory of general relativity. Here are some key contributions to scientific understanding facilitated by solar eclipses:

1. **Discovery of the Solar Corona:**
     - One of the most significant contributions of solar eclipses to science is the discovery and study of the solar corona, the Sun's outer atmosphere. During a total solar eclipse, the Moon blocks the Sun's bright disk, revealing the faint, pearly-white

corona surrounding it. Early astronomers used eclipses to observe and study the corona, leading to advancements in understanding its structure, dynamics, and behavior.

2. **Confirmation of General Relativity:**
   - In 1919, during a total solar eclipse, a team of astronomers led by Sir Arthur Eddington conducted observations to test Einstein's theory of general relativity. They observed the deflection of starlight passing near the Sun, as predicted by Einstein's theory, providing compelling evidence for its validity and revolutionizing our understanding of gravity.

3. **Discovery of Helium**:
   - Solar eclipses also played a role in the discovery of helium, an element first identified on the Sun. During a total solar eclipse in 1868, French astronomer

Jules Janssen and British astronomer Joseph Norman Lockyer independently observed a bright yellow emission line in the Sun's spectrum. They later identified this as the spectral signature of a previously unknown element, which they named helium after the Greek word for the Sun, "Helios."

4. **Advancements in Solar Physics:**
    - Solar eclipses have provided opportunities for scientists to study various phenomena associated with the Sun, including solar prominences, coronal mass ejections, and solar flares. Observations made during eclipses have contributed to our understanding of solar dynamics, magnetic fields, and the Sun-Earth connection.

5. **Atmospheric Science and Climate Studies:**
    - Solar eclipses offer unique opportunities to study atmospheric phenomena, such as changes in temperature, wind patterns, and cloud formation, during sudden changes in solar radiation. Scientists use data collected during eclipses to improve atmospheric models, study atmospheric composition, and investigate the impacts of solar variability on Earth's climate.
6. **Advancements in Imaging and Spectroscopy:**
    - Solar eclipses provide opportunities for astronomers to test new imaging techniques, instruments, and technologies for studying the Sun. High-resolution images and spectroscopic data obtained during eclipses contribute to

advancements in solar instrumentation and observational techniques.
7. **Public Engagement and Citizen Science:**
    - Solar eclipses inspire public interest and engagement in science, fostering opportunities for citizen science projects and outreach initiatives. Amateur astronomers and enthusiasts contribute to eclipse observations, data collection, and educational activities, enhancing our collective understanding of these celestial events.

By leveraging solar eclipses as natural laboratories for scientific inquiry, researchers continue to make significant strides in understanding the Sun, the Earth-Sun system, and the broader cosmos. These contributions underscore the enduring value of eclipses as opportunities for discovery, exploration, and collaboration in the pursuit of knowledge.

Notable Historical Eclipses

Throughout history, several solar eclipses have left an indelible mark on humanity, shaping scientific inquiry, cultural beliefs, and historical events. These notable eclipses have been documented and studied for their impact and significance. Here are a few examples:

1. **Thales' Eclipse (585 BCE):**
    - One of the earliest recorded solar eclipses occurred in 585 BCE and is attributed to the Greek philosopher Thales of Miletus. According to historical accounts, Thales predicted the eclipse, which temporarily halted a battle between the

Lydians and the Medes, leading to a peace agreement between the two warring factions.

2. **Solar Eclipse of 1919:**
   - The solar eclipse of 1919 gained renown for its role in confirming Albert Einstein's theory of general relativity. Expeditions led by Sir Arthur Eddington and others observed the deflection of starlight near the Sun during the eclipse, providing empirical evidence for Einstein's revolutionary theory of gravity.

3. **Solar Eclipse of 1878**:
   - The solar eclipse of 1878 traversed the western United States, drawing widespread attention and scientific interest. Thomas Edison used the occasion to test his newly invented tasimeter, a device designed to measure

temperature changes in the Sun's corona during an eclipse.

4. **Solar Eclipse of 1918:**
    - The solar eclipse of 1918, which crossed the United States from coast to coast, coincided with the height of World War I. Observations of the eclipse were coordinated by the National Academy of Sciences and contributed to advancements in solar physics and atmospheric science.

5. **Solar Eclipse of 1963:**
    - The solar eclipse of 1963, known as the "eclipse of the century," was visible across much of North America. It attracted millions of spectators and prompted extensive scientific observations, including studies of the solar corona and its magnetic field.

6. **Solar Eclipse of 1991:**
    - The solar eclipse of 1991, also known as the "Hawaii Eclipse," captivated observers in Hawaii and parts of Mexico. It provided scientists with opportunities to study the Sun's corona and chromosphere using advanced imaging techniques and instrumentation.
7. **Solar Eclipse of 2017:**
    - The solar eclipse of 2017, dubbed the "Great American Eclipse," traversed the contiguous United States from coast to coast, generating widespread excitement and media coverage. It was one of the most viewed eclipses in history, attracting millions of spectators to locations within the path of totality.

These notable historical eclipses represent just a fraction of the countless eclipses that have occurred throughout human history. Each eclipse has left its mark on science, culture, and society, contributing to our collective understanding of the cosmos and our place within it.

Cultural Significance

Myths, Legends, and Superstitions Surrounding Eclipses

Solar eclipses have long been shrouded in myths, legends, and superstitions across cultures and civilizations. These celestial events have inspired awe and wonder, as well as fear and uncertainty, leading to a rich tapestry of folklore and beliefs surrounding their occurrence. Here are some examples of myths, legends, and superstitions associated with eclipses:

1. **Mythological Battles:**
    - In various mythologies, solar eclipses are often depicted as cosmic battles between celestial

deities or supernatural forces. Ancient cultures interpreted the temporary darkening of the Sun as a conflict between light and darkness, good and evil, or divine beings vying for supremacy.

2. **Sun Devoured by Creatures:**
   - Some cultures believed that solar eclipses occurred when mythical creatures or celestial animals, such as dragons or serpents, devoured the Sun. People would engage in rituals or make loud noises to scare away these creatures and ensure the Sun's safe return.

3. **Divine Displeasure:**
   - Solar eclipses were sometimes interpreted as signs of divine displeasure or impending calamity. Ancient societies viewed eclipses as omens foretelling wars, natural disasters, or the downfall of

rulers. Rituals and sacrifices were performed to appease the gods and avert disaster.

4. **Temporary Darkness:**
   - In many cultures, the sudden darkness during a solar eclipse sparked fear and anxiety among the populace. People believed that malevolent spirits or supernatural entities roamed the Earth during eclipses, leading to customs such as staying indoors, covering wells, or refraining from eating.

5. **Taboos and Restrictions:**
   - Superstitions surrounding eclipses often led to taboos and restrictions on certain activities. For example, in some cultures, it was considered unlucky to eat or drink during an eclipse, as food and water might become

contaminated by the shadow's influence.

6. **Fertility and Pregnancy Beliefs:**
   - In some societies, eclipses were associated with fertility rites and beliefs about conception and childbirth. Pregnant women were often advised to avoid viewing eclipses or to take precautions to protect themselves and their unborn children from harm.

7. **Transformation and Renewal:**
   - Despite the fears and superstitions surrounding eclipses, many cultures also viewed them as symbols of transformation, renewal, and rebirth. Eclipses were seen as opportunities for personal or collective introspection, as well as for casting off old habits and embracing new beginnings.

While modern science has dispelled many of the myths and superstitions surrounding eclipses, these celestial events continue to evoke a sense of wonder and mystery. Today, solar eclipses serve as reminders of the enduring power of nature and our interconnectedness with the cosmos, inspiring curiosity, exploration, and cultural reflection.

Influence on Art, Literature, and Folklore

Solar eclipses have left an indelible mark on human culture, inspiring artists, writers, and storytellers throughout history. These celestial events have found expression in various forms of artistic expression, literature, and folklore, shaping cultural narratives and interpretations of the cosmos. Here are some ways in which solar eclipses have influenced art, literature, and folklore:

1. **Artistic Representation:**
   - Solar eclipses have been depicted in art across diverse cultures and time periods. Ancient cave paintings, medieval manuscripts, and Renaissance paintings often feature celestial motifs, including solar eclipses. Artists have captured the dramatic spectacle of eclipses, portraying the interplay of light and shadow, as well as the emotional impact on observers.
2. **Symbolism and Allegory:**
   - Solar eclipses are rich in symbolism and allegory, serving as potent metaphors for themes such as transformation, duality, and the passage of time. In literature and art, eclipses are often used to evoke a sense of mystery, awe, and existential contemplation, inviting viewers

to reflect on the cyclical nature of existence.

3. **Literary Inspiration:**
   - Solar eclipses have inspired writers and poets to explore themes of darkness and illumination, fate and free will, and the boundaries between the known and the unknown. From ancient myths and epic poems to modern novels and science fiction, eclipses feature prominently in literary works as catalysts for plot development, character arcs, and thematic exploration.

4. **Folklore and Mythology:**
   - Solar eclipses are steeped in folklore and mythology, with cultures around the world developing elaborate tales and beliefs to explain these celestial phenomena. Myths about solar eclipses often involve celestial deities, supernatural beings, and

epic battles between light and darkness. These stories serve as cultural touchstones, passing down wisdom, values, and traditions from generation to generation.

5. **Cultural Rituals and Ceremonies:**
    - Solar eclipses have inspired a wide range of cultural rituals, ceremonies, and customs designed to appease gods, ward off evil spirits, or harness the cosmic energies unleashed during an eclipse. From ancient rituals performed by indigenous peoples to modern-day eclipse festivals and celebrations, these cultural practices reflect humanity's enduring fascination with celestial events.

6. **Aesthetic Inspiration:**
    - The striking visual imagery of a solar eclipse—such as the corona's ethereal glow, the crescent-shaped Sun, and the

darkening of the sky—has inspired artists and designers to create evocative works of art, photography, and fashion. Eclipse-inspired designs often incorporate celestial motifs and cosmic themes, capturing the beauty and mystery of these celestial events.

Solar eclipses continue to inspire artistic expression, literary creativity, and cultural interpretation, serving as reminders of humanity's place in the cosmos and our enduring fascination with the mysteries of the universe. Through art, literature, and folklore, we explore the profound significance of eclipses as symbols of transcendence, transformation, and the eternal dance of light and shadow.

Eclipse Rituals and Traditions from Around the World

Solar eclipses have inspired a myriad of rituals and traditions in cultures across the globe, reflecting humanity's awe and reverence for these celestial events. From ancient ceremonies performed by indigenous peoples to modern-day practices rooted in cultural heritage, eclipse rituals offer a window into the diverse ways in which societies have sought to understand and interact with the cosmos. Here are some eclipse rituals and traditions from around the world:

1. **Navajo Nation (North America):**
    - The Navajo people of North America have sacred ceremonies and rituals associated with solar eclipses. Known as the "Serpent Chasing" ceremony, Navajo

elders and medicine men perform rituals to protect against negative influences and restore balance to the natural world during eclipses. Participants engage in prayer, chanting, and traditional dances to honor the Sun and seek blessings for the community.

2. **Hindu Traditions (India):**
    - In Hindu culture, eclipses are viewed as inauspicious events associated with negative energies and spiritual impurity. To counteract these influences, Hindus observe specific rituals known as "graha shanti," which involve fasting, bathing in holy rivers, and reciting prayers and mantras. Temples may also close during eclipses, and devotees are advised to avoid eating, drinking, or engaging in religious activities during the eclipse period.

3. **Inuit Practices (Arctic Regions):**
   - In Arctic regions inhabited by the Inuit and other indigenous peoples, solar eclipses are often seen as transformative events with spiritual significance. Inuit elders may share oral traditions and stories passed down through generations, highlighting the importance of maintaining harmony with nature and respecting the cycles of the Sun and Moon. Traditional practices such as drumming, singing, and storytelling may accompany eclipse observations.
4. **Japanese Beliefs (Japan):**
   - In Japanese folklore, solar eclipses are associated with mythical creatures and supernatural beings. Traditional beliefs hold that celestial dragons or demons known as "ryū" or "raiju" are responsible

for causing eclipses by swallowing the Sun. To ward off these malevolent spirits, people may engage in rituals such as clapping, banging pots, or making loud noises to frighten away the creatures and ensure the Sun's safe return.

5. **Maasai Customs (East Africa):**
    - Among the Maasai people of East Africa, solar eclipses are interpreted as omens or signs from the divine realm. Maasai elders may perform rituals and ceremonies to protect against negative influences and restore balance to the community. These rituals often involve chanting, dancing, and the sacrifice of livestock as offerings to appease the spirits and ensure the well-being of the tribe.
6. **Modern Observances (Global):**

- In modern times, solar eclipses continue to inspire awe and fascination, prompting people around the world to gather and witness these rare celestial events. Eclipse festivals, viewing parties, and educational events are organized in cities and towns along the path of totality, where enthusiasts come together to share in the experience of witnessing the spectacle of a total solar eclipse.

These eclipse rituals and traditions highlight the cultural diversity and spiritual significance of solar eclipses in societies around the world. Whether rooted in ancient mythology, religious beliefs, or communal practices, these rituals serve as reminders of humanity's enduring connection to the cosmos and our collective fascination with the mysteries of the universe.

Eclipse Viewing Safety

Importance of Eye Protection During Solar Eclipses

Ensuring proper eye protection during a solar eclipse is paramount to safeguarding one's vision and preventing potential eye damage. The intense brightness of the Sun, even when partially obscured during an eclipse, can cause serious harm to the eyes if viewed directly without adequate protection. Here are several reasons highlighting the importance of eye protection during solar eclipses:

1. **Risk of Solar Retinopathy:** Directly viewing the Sun, even for brief periods, can result in a condition known as solar retinopathy, where the Sun's focused rays damage the retina, the light-sensitive tissue at the back of the eye. Solar retinopathy can cause permanent vision loss or impairment, including blurred vision, blind spots, or distortion.
2. **Invisible Damage:** The danger of solar retinopathy lies in the fact that the Sun's harmful ultraviolet (UV) radiation and infrared (IR) radiation can cause damage to the retina without causing immediate pain or discomfort. This means that individuals may not realize they have injured their eyes until hours or days after exposure.
3. **Increased Risk During Eclipses:** During a solar eclipse, the temptation to look at the Sun is heightened due to the intriguing phenomenon of the Moon partially covering the Sun.

However, the danger to the eyes remains the same, if not greater, as individuals may be inclined to stare at the Sun for longer periods without adequate eye protection.

4. **Children and Vulnerable Populations:** Children are particularly susceptible to eye damage during solar eclipses, as they may be less aware of the risks and more likely to glance at the Sun out of curiosity. Additionally, individuals with certain eye conditions, such as macular degeneration or cataracts, may be at increased risk of eye damage from solar viewing.

5. **Permanent Consequences:** Unlike other forms of eye injury, solar retinopathy cannot be corrected through medical intervention or surgical procedures. Once the retina is damaged by solar exposure, the resulting vision loss or impairment may be permanent and irreversible.

6. **Proper Eye Protection:** The only safe way to view a solar eclipse is through the use of specialized solar viewing glasses or handheld solar filters specifically designed to block harmful UV and IR radiation. These glasses or filters must meet international safety standards (ISO 12312-2) to ensure adequate protection.

In summary, the importance of eye protection during solar eclipses cannot be overstated. Properly shielding one's eyes with certified solar viewing glasses or filters is essential to prevent irreversible eye damage and ensure a safe and enjoyable viewing experience of this extraordinary celestial event.

Safe Viewing Methods and Equipment for Solar Eclipses

To safely observe a solar eclipse and protect your eyes from potential damage, it's crucial to use proper viewing methods and equipment specifically designed for solar viewing. Here are some safe viewing methods and recommended equipment:

1. **Solar Viewing Glasses:**
    - Certified solar viewing glasses are specially designed to block harmful ultraviolet (UV) and infrared (IR) radiation emitted by the Sun. These glasses feature lenses that meet international safety standards (ISO 12312-2) and are labeled as safe for direct solar viewing. When using solar viewing glasses, ensure they are free from scratches, damage, or wear and tear.

2. **Handheld Solar Filters:**
   - Handheld solar filters, also known as solar filters or solar viewers, are another safe option for observing solar eclipses. These filters are typically made of aluminized Mylar or optical-grade solar film and can be attached to the front of telescopes, binoculars, or camera lenses to protect your eyes while viewing the Sun.
3. **Pinhole Projection:**
   - Pinhole projection is a simple and safe method for indirectly viewing a solar eclipse without looking directly at the Sun. To create a pinhole projector, poke a small hole in a piece of cardboard or paper and hold it up to the Sun, allowing the sunlight to pass through the pinhole and project an image of the eclipsed Sun onto a surface,

such as a piece of white paper or the ground.
4. **Telescopes with Solar Filters:**
    - Telescopes equipped with solar filters provide a magnified view of the Sun during a solar eclipse, allowing observers to see intricate details such as sunspots, solar flares, and the solar corona. Ensure that the telescope's solar filter is securely attached to the front aperture and is specifically designed for solar viewing to prevent eye damage.
5. **Binoculars with Solar Filters:**
    - Binoculars fitted with solar filters offer a close-up view of the Sun's surface and features, similar to telescopes. Use caution when using binoculars for solar viewing and ensure that the solar filters are securely attached to both eyepieces to

protect your eyes from harmful solar radiation.

6. **Digital Cameras with Solar Filters:**
   - Digital cameras equipped with solar filters allow photographers to capture stunning images of solar eclipses without risking eye damage. Attach a solar filter to the camera lens to block harmful UV and IR radiation while capturing high-resolution photos or videos of the eclipse.

7. **Online Livestreams and Broadcasts:**
   - If viewing a solar eclipse directly is not possible or safe, consider watching live streams or broadcasts of the eclipse online or on television. Many scientific organizations, observatories, and media outlets provide real-time coverage of solar eclipses, allowing viewers to experience

the event from the safety and comfort of their homes.

Remember to always exercise caution and follow safety guidelines when observing a solar eclipse. Never look directly at the Sun without proper eye protection, as doing so can cause permanent eye damage or blindness. With the right equipment and viewing methods, you can safely enjoy the breathtaking spectacle of a solar eclipse while protecting your vision for years to come.

Health risks of viewing eclipses improperly

Viewing a solar eclipse improperly can pose serious health risks to your eyesight. Here are some of the potential health risks associated with viewing eclipses without proper eye protection:

1. **Solar Retinopathy:** Directly viewing the Sun during a solar eclipse, even for a short period, can result in solar retinopathy. This condition occurs when the intense light from the Sun damages the cells of the retina, the light-sensitive tissue at the back of the eye. Solar retinopathy can cause permanent vision loss or impairment, including blurred vision, blind spots, or distorted vision.
2. **Photokeratitis**: Exposure to the Sun's ultraviolet (UV) radiation during a solar eclipse can cause photokeratitis, also known as "snow blindness" or "welder's flash." Photokeratitis is characterized by inflammation of the cornea, the clear outer layer of the eye, resulting in symptoms such as pain, redness, tearing, and sensitivity to light. While photokeratitis is usually temporary, repeated exposure to UV radiation without protection can increase the risk of long-term eye damage.

3. **Macular Damage:** The central part of the retina, known as the macula, is particularly vulnerable to damage from solar radiation. Prolonged or intense exposure to the Sun's bright light during a solar eclipse can lead to macular damage, causing central vision loss or impairment. Macular damage can significantly impact daily activities such as reading, driving, and recognizing faces.
4. **Delayed Symptoms:** One of the dangers of viewing a solar eclipse improperly is that the damage to the eyes may not be immediately apparent. Solar retinopathy and other forms of eye damage can have delayed onset symptoms, meaning that individuals may not realize they have injured their eyes until hours or days after exposure. By then, irreversible damage may have already occurred.
5. **Increased Risk for Children:** Children are particularly vulnerable to

eye damage during solar eclipses, as they may be less aware of the risks and more likely to glance at the Sun out of curiosity. Parents and caregivers should closely supervise children during eclipses and ensure they have proper eye protection if they wish to view the event safely.
6. **Permanent Consequences:** Unlike other forms of eye injury, solar retinopathy and macular damage cannot be corrected through medical intervention or surgical procedures. Once the delicate cells of the retina are damaged by solar radiation, the resulting vision loss or impairment may be permanent and irreversible.

To protect your eyesight and avoid these potential health risks, it's essential to use certified solar viewing glasses, handheld solar filters, or other safe viewing methods when observing a solar eclipse. Never look directly at the Sun without proper eye protection, as doing so can have serious and long-lasting consequences for your vision.

Planning for an Eclipse

How to prepare for an eclipse viewing

Preparing for a solar eclipse viewing involves several important steps to ensure a safe and enjoyable experience. Here's a guide on how to prepare for an eclipse viewing:

1. **Research the Eclipse:**
    - Familiarize yourself with the date, time, and duration of the solar eclipse you plan to observe. Research the path of totality and determine if you will be within the optimal

viewing zone to experience the full spectacle of totality.

2. **Choose a Viewing Location:**
   - Select a suitable location with an unobstructed view of the sky, free from tall buildings, trees, or other obstacles that may block your view of the Sun during the eclipse. Consider factors such as accessibility, weather conditions, and proximity to amenities.

3. **Check Weather Forecasts:**
   - Monitor weather forecasts for your chosen viewing location in the days leading up to the eclipse. Clear skies are essential for optimal viewing, so be prepared to adjust your plans if inclement weather is forecasted.

4. **Obtain Certified Eye Protection:**
   - Purchase certified solar viewing glasses or handheld solar filters from reputable vendors. Ensure

that the eye protection you choose meets international safety standards (ISO 12312-2) and is labeled as safe for direct solar viewing. Check the glasses or filters for any signs of damage or wear before use.

5. **Gather Viewing Equipment:**
    - If you plan to use telescopes, binoculars, or cameras for solar viewing, ensure that they are equipped with proper solar filters to protect your eyes from harmful solar radiation. Test your equipment beforehand to familiarize yourself with its operation and ensure that it is functioning correctly.

6. **Plan for Transportation and Logistics:**
    - If you need to travel to reach your chosen viewing location, plan your transportation and logistics well in advance. Consider factors such as traffic

congestion, parking availability, and road closures, especially if you anticipate large crowds converging on popular viewing sites.

7. **Pack Essentials and Comfort Items:**
   - Bring along essential items such as water, snacks, sunscreen, hats, and comfortable clothing to ensure your comfort during the eclipse viewing. Consider packing a portable chair or blanket for seating and a portable shade canopy or umbrella for sun protection.

8. **Educate Yourself and Others:**
   - Take the time to educate yourself and others about the importance of proper eye protection and safe viewing practices during a solar eclipse. Share information about the risks of looking directly at the Sun without protection and

encourage others to use certified solar viewing glasses or filters.

9. **Stay Informed and Flexible:**
    o Stay informed about any updates or changes to eclipse viewing conditions, including weather updates, traffic advisories, and safety guidelines. Be flexible and prepared to adapt your plans as needed to ensure a safe and enjoyable eclipse viewing experience.

By following these steps and taking proper precautions, you can prepare yourself for an eclipse viewing experience that is both safe and memorable. Remember to prioritize your safety and the safety of others while enjoying the awe-inspiring spectacle of a solar eclipse.

Choosing a viewing location

When choosing a viewing location for a solar eclipse, consider several factors to ensure an optimal and memorable experience. Here are some tips for selecting the perfect viewing spot:

1. **Path of Totality:**
    - If possible, choose a location within the path of totality, where the Moon will completely cover the Sun, resulting in a total solar eclipse. Being within the path of totality offers the most dramatic and immersive viewing experience, with the sky darkening and the Sun's corona visible to the naked eye.
2. **Accessibility**:
    - Select a viewing location that is easily accessible by car, public transportation, or on foot. Consider factors such as

parking availability, proximity to major roads or highways, and ease of navigation to and from the site.
3. **Unobstructed View:**
   - Look for a location with an unobstructed view of the sky, free from tall buildings, trees, or other structures that may block your view of the Sun during the eclipse. Open fields, parks, beaches, and hilltops are ideal for uninterrupted viewing.
4. **Safety and Comfort:**
   - Prioritize safety and comfort when choosing a viewing location. Select a site away from busy roads or crowded areas to minimize distractions and ensure a safe environment for observing the eclipse. Bring along essential items such as water, sunscreen, hats, and seating for added comfort during the viewing experience.

5. **Weather Conditions:**
   - Monitor weather forecasts for your chosen viewing location in the days leading up to the eclipse. Clear skies are essential for optimal viewing, so select a location with favorable weather conditions and be prepared to adjust your plans if inclement weather is forecasted.

6. **Crowd Considerations:**
   - Be mindful of potential crowds and congestion at popular viewing sites, especially within the path of totality. Consider arriving early to secure a good viewing spot and avoid last-minute rush or traffic delays.

7. **Amenities and Facilities:**
   - Choose a location with access to amenities such as restrooms, food vendors, and shaded areas, especially if you plan to spend

an extended period observing the eclipse. Public parks, campgrounds, and recreational areas may offer convenient facilities for eclipse viewers.

8. **Scenic Beauty:**
    - Enhance your eclipse viewing experience by selecting a location with scenic beauty or natural attractions. Choose sites with panoramic views, picturesque landscapes, or iconic landmarks to complement the awe-inspiring spectacle of the solar eclipse.

9. **Local Regulations and Permissions:**
    - Familiarize yourself with any local regulations or permissions governing eclipse viewing at your chosen location. Some sites may have restrictions on parking, camping, or public gatherings, so be sure to follow rules and guidelines to ensure a

safe and enjoyable experience for everyone.

By considering these factors and choosing a viewing location that meets your preferences and needs, you can maximize your chances of witnessing a spectacular solar eclipse and creating lasting memories of this extraordinary celestial event.

Weather considerations

When planning for a solar eclipse viewing, weather considerations are crucial for ensuring optimal visibility and a memorable experience. Here are some important weather factors to consider:

1. **Clear Skies:**
    - Clear skies are essential for observing a solar eclipse. Check weather forecasts for

your chosen viewing location in the days leading up to the eclipse to assess the likelihood of clear skies during the event. Look for locations with low cloud cover and minimal atmospheric haze for optimal viewing conditions.

2. **Cloud Cover:**
   - Monitor cloud cover forecasts for your viewing area to gauge the likelihood of obstructed views during the eclipse. Choose locations with low cloud cover probabilities to maximize your chances of seeing the eclipse clearly. Be prepared to adjust your plans or relocate to an alternate viewing site if clouds are forecasted to obscure the Sun.
3. **Rainfall**:
   - Rainfall can significantly impact eclipse visibility and viewing conditions. Avoid

locations with high chances of precipitation during the eclipse timeframe, as rain clouds can obscure the Sun and hinder observation. Consider indoor viewing options or alternate activities in case of inclement weather.

4. **Wind Conditions:**
    - Windy conditions can affect stability and comfort during eclipse viewing, especially if using telescopes, cameras, or other equipment. Choose sheltered locations or areas with natural windbreaks, such as trees or buildings, to minimize the impact of strong winds on your viewing experience.
5. **Temperature**:
    - Consider the temperature forecast for your viewing location and dress accordingly to ensure comfort during the

eclipse. Wear layers to adapt to changing temperatures throughout the event and bring extra clothing or blankets if viewing during cooler hours, such as early morning or late afternoon.

6. **Heat and Sun Exposure:**
    - If viewing the eclipse during peak daylight hours, take precautions to protect yourself from excessive heat and sun exposure. Wear sunscreen, sunglasses, hats, and lightweight, breathable clothing to shield yourself from the Sun's intense rays and prevent sunburn or heat-related illnesses.

7. **Safety Considerations:**
    - Be mindful of safety hazards associated with adverse weather conditions, such as lightning, thunderstorms, or high winds. Seek shelter indoors or in a

sturdy, enclosed structure if severe weather threatens your safety during the eclipse. Avoid open fields, exposed areas, or water bodies during electrical storms.

8. **Flexibility and Backup Plans:**
    - Stay flexible and prepared to adjust your eclipse viewing plans based on changing weather conditions and forecasts. Have backup options or alternate viewing sites in mind in case your initial location is affected by poor weather. Consider indoor viewing opportunities or live stream broadcasts as alternative options if outdoor viewing is not feasible.

By carefully considering weather forecasts and conditions, you can increase your chances of experiencing a clear and unobstructed view of the solar eclipse while ensuring your safety and comfort throughout the event.

Experiencing a Solar Eclipse

What to expect during each phase of the eclipse

During a solar eclipse, several distinct phases occur as the Moon passes between the Earth and the Sun, resulting in various stages of partial and total eclipse. Here's what to expect during each phase of a typical solar eclipse:

1. **Partial Eclipse:**
    - The partial eclipse phase begins when the Moon first begins to move across the face of the Sun, partially blocking its light. At the beginning of this phase, a small "bite" or crescent shape

appears to be taken out of the Sun's disk. As the eclipse progresses, the Moon gradually covers more of the Sun, darkening the sky and creating a noticeable decrease in ambient light.

2. **Progression of Partiality:**
    - During the partial eclipse phase, the Moon continues to move across the Sun, gradually covering a larger portion of its disk. The degree of partiality varies depending on the observer's location within the eclipse path, with some areas experiencing greater coverage than others. Observers may notice changes in lighting conditions, temperature, and wildlife behavior as the eclipse progresses.
3. **Approach to Totality (For Total Eclipses):**

- In the moments leading up to totality during a total solar eclipse, observers may experience a phenomenon known as the "diamond ring effect" or "Baily's beads." As the Moon's disk almost completely covers the Sun, a brief flash of light resembling a sparkling diamond ring appears on the edge of the lunar silhouette, accompanied by a series of bright spots or "beads" caused by sunlight streaming through lunar valleys and craters.

4. **Totality (For Total Eclipses):**
    - Totality marks the peak phase of a total solar eclipse, occurring when the Moon completely obscures the Sun, plunging the surrounding area into darkness. During totality, observers witness the Sun's corona, the outer atmosphere of

the Sun, glowing in a faint, ethereal halo around the darkened disk of the Moon. The sky may take on surreal hues, and stars and planets become visible in the daytime sky.

5. **Totality Duration and Eclipse Path:**
   - The duration of totality varies depending on the observer's location within the eclipse path, with totality lasting for a few seconds to several minutes. Observers situated near the centerline of the eclipse path experience the longest duration of totality, while those on the outskirts may experience shorter durations. Eclipse chasers often travel to prime viewing locations to maximize their time in totality.

6. **Return to Partiality:**
   - After totality ends, the partial eclipse phase resumes as the Moon begins to move away

from the Sun's disk, gradually uncovering its light. The process of partiality reversal mirrors the progression of the partial eclipse phase in reverse, with the Sun's crescent shape gradually expanding as the Moon moves further away.

7. **Conclusion of the Eclipse:**
    o The eclipse concludes when the Moon no longer blocks any part of the Sun's disk, marking the end of the partial eclipse phase. Observers may experience a return to normal lighting conditions and ambient temperatures as the eclipse reaches its conclusion.

By understanding the sequence of phases during a solar eclipse, observers can anticipate and appreciate the gradual changes in the sky and lighting conditions throughout the event, culminating in the awe-inspiring spectacle of totality during a total solar eclipse.

Emotional and psychological impact of totality

The experience of totality during a solar eclipse can have a profound emotional and psychological impact on observers, eliciting a wide range of reactions and responses. Here are some aspects of the emotional and psychological impact of totality:

1. **Sense of Awe and Wonder:**
    - Witnessing the sudden darkness and the appearance of the Sun's corona during totality can evoke a profound sense of awe and wonder. Observers often describe feeling humbled by the majesty of the cosmos and the beauty of the celestial spectacle unfolding before them.
2. **Overwhelming Emotions:**

- Many observers report experiencing intense emotions during totality, ranging from excitement and euphoria to awe and reverence. The sheer magnitude of the event and the fleeting nature of totality can evoke powerful emotional responses that linger long after the eclipse has ended.

3. **Connection to Nature and the Universe:**
    - Totality provides a rare opportunity to feel connected to the natural world and the vastness of the universe. Observers often describe feeling a sense of unity with the cosmos and a deeper appreciation for the interconnectedness of all living things.

4. **Sense of Transcendence:**
    - For some observers, witnessing totality can be a transcendent

experience, momentarily transcending mundane concerns and everyday worries. The surreal beauty of the eclipsed Sun and the darkened sky can create a sense of timelessness and spiritual resonance.

5. **Communal Bonding:**
    - Eclipse viewing often fosters a sense of communal bonding and shared experience among observers. Whether watching with friends, family, or strangers, the collective witnessing of totality can create lasting memories and forge connections with others who share a common fascination with the cosmos.

6. **Personal Transformation:**
    - The experience of totality has the potential to spark personal transformation and introspection. Observers may find themselves reflecting on

the fleeting nature of life, the passage of time, and the profound mysteries of the universe, leading to moments of self-discovery and existential contemplation.

7. **Post-Eclipse Withdrawal:**
    - After experiencing totality, some observers may undergo a period of post-eclipse withdrawal, characterized by feelings of longing or nostalgia for the fleeting moments of darkness and celestial beauty. This phenomenon, known as "eclipse withdrawal," highlights the profound impact that totality can have on individuals' emotional and psychological well-being.

Overall, the emotional and psychological impact of totality during a solar eclipse is a deeply personal and subjective experience, shaped by individual perspectives, beliefs, and prior experiences. For many observers, witnessing totality is a transformative and unforgettable moment that leaves a lasting imprint on their hearts and minds, inspiring a lifelong fascination with the wonders of the universe.

Capturing eclipse phenomena through photography

Capturing eclipse phenomena through photography requires careful planning, the right equipment, and technical know-how to ensure stunning and memorable images. Here are some tips for photographing eclipse phenomena:

1. **Use Proper Equipment:**
   - Invest in a DSLR or mirrorless camera with manual exposure controls, which allows for greater flexibility and control over your photography settings. Attach a telephoto lens or zoom

lens with a focal length of at least 200mm to capture detailed images of the Sun and eclipse phenomena.

2. **Solar Filters:**
    - Never photograph the Sun without using proper solar filters to protect your eyes and camera equipment from harmful solar radiation. Use certified solar filters designed specifically for photography to block out most of the Sun's intense light and heat. Attach the solar filter securely to the front of your lens to prevent any light leakage.
3. **Tripod Stability:**
    - Use a sturdy tripod to keep your camera stable and minimize camera shake during long exposures. This is especially important when using telephoto lenses or

shooting at slow shutter speeds to capture eclipse phenomena.
4. **Manual Settings:**
    - Switch your camera to manual mode to have full control over exposure settings. Set your aperture to a narrow f-stop (e.g., f/8 to f/16) to ensure sharpness and depth of field. Adjust the shutter speed and ISO sensitivity based on the available light conditions and desired exposure.
5. **Bracketing Exposures:**
    - Consider bracketing your exposures by taking multiple shots at different exposure settings (e.g., varying shutter speeds or aperture sizes). This technique helps ensure that you capture the full range of brightness and detail in the eclipse phenomena, from the bright solar disk to the faint corona.

6. **Focus Accuracy:**
    - Achieve precise focus by using manual focus mode and magnifying the Sun's image in your camera's viewfinder or LCD screen. Focus on a distinct feature of the Sun, such as sunspots or the edge of the solar disk, to ensure sharpness in your images.
7. **Composition and Framing:**
    - Experiment with composition and framing to create visually compelling images of the eclipse phenomena. Consider including elements of the surrounding landscape or silhouetted foreground objects to add context and interest to your photographs.

8. **Capture Eclipse Phases:**
    - Document the various phases of the eclipse, from the partial

phases to totality (if applicable), to create a comprehensive visual narrative of the event. Take multiple shots throughout the eclipse progression to capture the changing appearance of the Sun and surrounding environment.

9. **Long Exposures and Time-Lapse:**
    - Experiment with long exposures or time-lapse photography to capture the dynamic motion of the eclipse phenomena, such as the movement of the Moon's shadow across the landscape or the gradual darkening of the sky during totality. Use a remote shutter release or intervalometer to trigger your camera's shutter at regular intervals.

10. **Practice and Patience:**
    - Practice your photography techniques and familiarize

yourself with your equipment before the day of the eclipse. Be patient and persistent, as capturing stunning eclipse images may require multiple attempts and adjustments to achieve the desired results.

By following these tips and techniques, you can capture breathtaking and memorable images of eclipse phenomena, from the partial phases to the awe-inspiring spectacle of totality, preserving the beauty and wonder of these celestial events for years to come.

Beyond the Eclipse

Scientific research during eclipses

Solar eclipses provide unique opportunities for scientific research across various disciplines, allowing scientists to study phenomena that are not readily observable under normal conditions. Here are some examples of scientific research conducted during eclipses:

1. **Solar Physics:**
    - Solar eclipses offer scientists the chance to study the Sun's outer atmosphere, known as the corona, which is visible as a faint halo of light during totality. Researchers use specialized instruments, such as

coronagraphs and spectrometers, to observe the corona's structure, temperature, and magnetic fields, providing insights into solar dynamics and space weather.

2. **Stellar and Planetary Atmospheres:**
   - By observing the gradual dimming and reddening of stars and planets during a solar eclipse, astronomers can study the properties of their atmospheres, including temperature, composition, and density. This information helps researchers better understand the atmospheres of distant celestial bodies and refine models of planetary and stellar atmospheres.

3. **Earth's Atmosphere and Climate:**
   - During a solar eclipse, changes in temperature, humidity, and atmospheric pressure occur as the Moon blocks out the Sun's

radiation. Scientists use ground-based instruments and satellite data to monitor these atmospheric changes and their effects on weather patterns, cloud formation, and atmospheric circulation. Studying eclipse-induced atmospheric phenomena contributes to our understanding of Earth's climate system.

4. **Biological and Ecological Research:**
   - Solar eclipses provide opportunities to study the effects of sudden changes in light levels on plants, animals, and ecosystems. Researchers investigate how eclipse-induced darkness influences behavior, physiology, and ecology in various species, shedding light on the adaptive strategies of living organisms to environmental fluctuations.

5. **Space and Astrophysics:**
    - Observations of solar eclipses from high-altitude aircraft, balloons, or spacecraft enable scientists to study phenomena such as the solar corona, solar prominences, and the solar wind in greater detail. Instruments aboard these platforms capture images and data that complement ground-based observations, enhancing our understanding of solar and space physics.
6. **Technology Testing and Calibration:**
    - Solar eclipses provide opportunities to test and calibrate scientific instruments, telescopes, and imaging systems under challenging observational conditions. Scientists use eclipses as calibration events to validate instrument performance, refine

measurement techniques, and improve data processing algorithms for future scientific missions.
7. **Educational Outreach and Public Engagement:**
    - Solar eclipses capture public interest and imagination, making them valuable opportunities for educational outreach and public engagement in science. Scientists collaborate with educators, museums, and outreach organizations to develop educational programs, citizen science projects, and public viewing events that promote scientific literacy and inspire the next generation of researchers.

By leveraging the unique opportunities presented by solar eclipses, scientists can advance knowledge and understanding across a wide range of scientific disciplines, contributing to discoveries and breakthroughs that benefit society as a whole.

Citizen science opportunities

Solar eclipses present exciting opportunities for citizen science participation, allowing individuals of all ages and backgrounds to contribute valuable data and observations to scientific research. Here are some citizen science opportunities associated with solar eclipses:

1. **Eclipse Megamovie Project:**
    - The Eclipse Megamovie Project engages citizen scientists in capturing images and videos of total solar eclipses from multiple locations along the eclipse path. Participants use smartphones or digital cameras to document the eclipse and submit their observations to the

project's online platform. The collected imagery is then compiled into a "mega movie" that provides a comprehensive view of the eclipse from various vantage points, aiding scientific research on solar phenomena and atmospheric dynamics.

2. **GLOBE Observer:**
   - GLOBE Observer is a citizen science program that encourages participants to contribute environmental observations to NASA's Global Learning and Observations to Benefit the Environment (GLOBE) program. During solar eclipses, participants can collect data on air temperature, cloud cover, and other atmospheric conditions using the GLOBE Observer app. These observations help scientists study the effects of

eclipses on the Earth's atmosphere and climate.
3. **Solar Eclipse Citizen Science Projects:**
    - Various research institutions and organizations host citizen science projects focused specifically on solar eclipses. These projects may involve collecting data on animal behavior, temperature changes, cloud cover, or other environmental factors during the eclipse. Participants can contribute observations, measurements, and photographs to help scientists better understand the impacts of eclipses on the natural world.

4. **Planetarium and Museum Programs:**
    - Planetariums, museums, and science centers often organize

citizen science events and activities in conjunction with solar eclipses. These programs may include hands-on workshops, educational presentations, and public outreach events where participants can learn about eclipse science and contribute to ongoing research projects. Citizen scientists can engage in data collection, analysis, and interpretation under the guidance of expert researchers and educators.

5. **Educational Outreach Initiatives:**
   - Solar eclipses provide excellent opportunities for educational outreach initiatives aimed at engaging students, teachers, and the general public in scientific inquiry. Citizen science projects associated with eclipses may be incorporated into K-12 curriculum materials,

informal science learning programs, and community science events, allowing participants to learn about eclipse phenomena while making meaningful contributions to scientific research.

By participating in citizen science projects during solar eclipses, individuals can actively contribute to scientific discovery, enhance their understanding of eclipse phenomena, and become part of a global community of eclipse enthusiasts and researchers. Citizen science initiatives not only advance scientific knowledge but also foster public engagement in science and promote lifelong learning opportunities for people of all ages.

Future eclipse predictions and events

Looking ahead, several future solar eclipses are predicted to occur in different parts of the world, providing opportunities for observation and scientific study. Here are some upcoming eclipse predictions and events:

1. **April 8, 2024: Total Solar Eclipse:**
   - The next total solar eclipse is set to take place on April 8, 2024. Dubbed the "Great North American Eclipse," this event will be visible across North America, with the path of totality stretching from Mexico through the United States and up into Canada. Major cities within the path of totality include Dallas, Indianapolis, Cleveland, Buffalo, and Montreal.

2. **August 12, 2026: Total Solar Eclipse:**
    - Another total solar eclipse is forecasted to occur on August 12, 2026. This eclipse will be visible from parts of North America, Europe, and Africa. The path of totality will pass over the Arctic regions, including Greenland, Iceland, and parts of Canada and Russia.
3. **August 2, 2027: Total Solar Eclipse:**
    - A total solar eclipse is predicted to occur on August 2, 2027. This eclipse will be visible from parts of North Africa, the Middle East, and Central Asia. The path of totality will traverse regions including Morocco, Spain, Algeria, Libya, Egypt, Saudi Arabia, and Yemen.
4. **June 25, 2030: Annular Solar Eclipse:**

- An annular solar eclipse is anticipated to take place on June 25, 2030. This eclipse will be visible from parts of Europe, Asia, and Africa. The path of annularity will pass over regions such as Spain, France, Italy, Greece, Turkey, Russia, and China.

These are just a few examples of future solar eclipses that astronomers and enthusiasts can look forward to observing and studying. Solar eclipses continue to captivate people around the world, inspiring scientific research, cultural traditions, and public fascination with the wonders of the cosmos. As technology advances, astronomers and researchers will be able to make increasingly accurate predictions of eclipse events, providing opportunities for observation and discovery for generations to come.

Frequently Asked Questions

common misconceptions and concerns

Addressing common misconceptions and concerns about solar eclipses is essential for promoting accurate information and ensuring public safety during these celestial events. Here are some common misconceptions and concerns, along with clarifications:

1. **Myth: Solar eclipses are rare occurrences.**
    - Clarification: While total solar eclipses may seem rare from any specific location on Earth, they occur approximately every 18 months somewhere on the planet. However, the path of

totality for each eclipse is relatively narrow, making it less likely for any one location to experience totality frequently.
2. **Myth**: It's safe to look at the Sun during a solar eclipse without eye protection.
    - Clarification: Looking directly at the Sun, even during an eclipse, can cause permanent eye damage or blindness. It's crucial to use certified solar viewing glasses or handheld solar filters to protect your eyes when observing a solar eclipse. Regular sunglasses are not sufficient for safe solar viewing.
3. **Concern**: Animals may behave strangely or become agitated during a solar eclipse.
    - Clarification: While some anecdotal reports suggest changes in animal behavior during eclipses, scientific

studies have not consistently confirmed these observations. It's essential to treat animals with care and observe safety precautions during eclipses, but there's no need for undue concern about their behavior.

4. **Myth**: Solar eclipses have harmful effects on pregnant women or unborn babies.
    - Clarification: There is no scientific evidence to suggest that solar eclipses pose any specific risks to pregnant women or their babies. However, pregnant women should still take the same precautions as others when viewing eclipses, such as using proper eye protection and avoiding looking directly at the Sun.
5. **Concern**: Traffic congestion and overcrowding may occur at eclipse viewing sites.

- Clarification: It's common for popular eclipse viewing locations to experience increased traffic and crowds as enthusiasts gather to witness the event. Planning ahead, arriving early, and considering alternative viewing sites can help mitigate congestion and ensure a smoother experience for eclipse viewers.

6. **Myth**: Solar eclipses are harbingers of doom or bring bad luck.
    - Clarification: Solar eclipses have been surrounded by myths, superstitions, and cultural beliefs throughout history. However, from a scientific standpoint, eclipses are natural phenomena caused by the Moon's orbit intersecting with the Earth and the Sun. There is no evidence to support claims of negative effects associated with eclipses.

7. **Concern**: Watching a solar eclipse through unfiltered lenses, such as camera or telescope lenses, is safe.
   - Clarification: Looking at the Sun through unfiltered lenses, such as camera or telescope lenses, can cause severe eye damage or blindness. Always use proper solar filters designed for direct solar viewing to protect your eyes and equipment during eclipses.

By addressing these common misconceptions and concerns with accurate information and safety recommendations, individuals can enjoy solar eclipses safely and responsibly, while appreciating the awe-inspiring beauty of these celestial events.

Answers to common questions about eclipses

1. **What is a solar eclipse?**
   - A solar eclipse occurs when the Moon passes between the Earth and the Sun, blocking all or part of the Sun's light from reaching Earth.
2. **What causes a solar eclipse?**
   - A solar eclipse occurs when the Moon's shadow falls on Earth's surface, blocking the Sun's light. This happens when the Moon is in its new moon phase and aligns directly between the Earth and the Sun.
3. **What is the difference between a total, partial, and annular solar eclipse?**
   - In a total solar eclipse, the Moon completely covers the Sun, turning day into darkness for a brief period. In a partial

solar eclipse, only part of the Sun is obscured by the Moon. An annular solar eclipse occurs when the Moon is too far from Earth to completely cover the Sun, resulting in a ring of sunlight around the Moon's silhouette.

4. **How often do solar eclipses occur?**
    - Solar eclipses occur approximately every 18 months, but the path of totality for each eclipse is relatively narrow, making it less likely for any one location to experience totality frequently.

5. **Is it safe to look at a solar eclipse?**
    - No, it is never safe to look directly at the Sun without proper eye protection, even during a solar eclipse. Doing so can cause permanent eye damage or blindness. Always use certified solar viewing glasses or handheld solar filters

to protect your eyes when observing a solar eclipse.

6. **Where is the best place to view a solar eclipse?**
    - The best place to view a solar eclipse depends on the specific path of totality for that eclipse. Eclipse chasers often travel to locations within the path of totality to experience the full spectacle of totality, where the Sun is completely obscured by the Moon.

7. **What should I do if I can't see the eclipse in person?**
    - If you are unable to see the eclipse in person, you can still enjoy the event through live streams, webcasts, and broadcasts provided by observatories, science centers, and media outlets. Many organizations offer online resources and virtual

experiences to engage with eclipses remotely.

8. **Can animals sense solar eclipses?**
   - While some anecdotal reports suggest changes in animal behavior during eclipses, scientific studies have not consistently confirmed these observations. It's essential to treat animals with care and observe safety precautions during eclipses, but there's no need for undue concern about their behavior.

9. **How can I photograph a solar eclipse safely?**
   - Photographing a solar eclipse safely requires using proper solar filters to protect both your eyes and your camera equipment. Never look at the Sun through an unfiltered camera lens or viewfinder. Follow recommended techniques for solar

photography and use certified solar viewing glasses for eye protection.

10. When is the next solar eclipse?
- The next solar eclipse will occur on August 12, 2026. The path of totality for this eclipse will span North America, Europe, and Africa, providing an opportunity for observers within the path to experience the full spectacle of totality.

Thanks! Readers

Dear Reader,

Thank you for choosing to embark on this journey into the captivating world of solar eclipses with "Solar Eclipse 2024." Your support and interest in exploring the wonders of the cosmos are truly appreciated.

As you delve into the pages of this book, we hope you find yourself immersed in the science, myths, and cultural significance of solar eclipses, as well as equipped with practical guidance for safe observation and photography techniques.

Your feedback and thoughts on the book are invaluable to us. If you enjoyed the experience and found the information helpful, we kindly ask you to consider leaving a review to share your thoughts with others. Your review can help guide fellow enthusiasts and curious minds in their quest for knowledge about solar eclipses.

Looking ahead to future celestial events, we encourage you to apply the insights and instructions provided in this book during the next solar eclipse and any other opportune moments for stargazing and astronomical observation. By following the recommended safety precautions and techniques, you can enhance your eclipse viewing experience while safeguarding your eyes and equipment.

Thank you once again for joining us on this celestial journey. May your curiosity continue to be ignited by the mysteries of the universe, and may you always find inspiration in the beauty of the cosmos.

Happy exploring!

Sincerely,
Mark M. Fields

www.ingramcontent.com/pod-product-compliance
Lightning Source LLC
Chambersburg PA
CBHW070145230526
45471CB00002B/517